CONTENTS

WHAT IS AI?

Artificial Intelligence (AI) is the science, technology and engineering of intelligent machines. AI is all around you – a part of your everyday life. A lot of the time you might not even realise that the devices you interact with are 'intelligent'!

AMAZING AI ABILITIES

The goal of AI is to create machines that use human-like intelligence to perform many different tasks. These might be practical jobs, such as mowing the lawn. But they may also be incredibly tricky tasks where the machine has to think and learn.

HUMANS – THE ULTIMATE MACHINES

Your brain is an amazing machine, carrying out complex processes every second of every day. It's how you think, feel, react, reason, analyse, learn and explain. Understanding human abilities like these is key to AI – recreating these processes in machines is what makes artificial intelligence so 'real'.

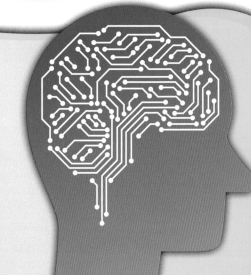

WHAT ARE SMART DEVICES?

Smart devices are everyday electronic objects that use computers as a kind of 'brain' to fuel their intelligence. Smart devices are often small – but they're powerful, too! They usually connect with other devices to collect and share **data**. Smart devices include:

PHONES

TABLETS

SPEAKERS

WATCHES

THERMOSTATS

WHERE WILL IT LEAD?

As scientists and engineers work to build devices that predict and respond to our needs more effectively, some people have started to wonder whether all this smart technology is a good thing. AI engineers consider **ethical** questions carefully in their work.

?

Just because we *can* develop smarter devices, does that mean we *should*?

 Do we really need them?

 Will we start to rely on them too much?

 Who decides what 'too much' is?

5

WORK, WORK, WORK

Smart devices come in many different shapes and forms. But they are mostly designed with one goal in mind: to make our lives simpler and easier. Smart devices take on tasks that we would otherwise have to do ourselves.

WEAK AND STRONG AI

Experts divide AI into two broad types: strong AI and weak (or narrow) AI.

Strong AI refers to intelligence that can think and learn on its own. This kind of technology can carry out multiple tasks.

Weak AI refers to AI technology that is geared towards one particular task. Most smart devices are an example of weak AI.

CLEAN AND TIDY

Robotic vacuum cleaners and lawnmowers are two of the most popular types of practical work-related smart devices. They use weak AI to make decisions based on information they gather through **sensors**. For example, they can sense the size of a room or garden, and can detect and avoid obstacles. They can also find their way back to a charging point when their batteries are low.

Some smart vacuum cleaners can even remember the most efficient route around a room to get the cleaning done!

SMART FACTORIES

You've probably heard of a smart home (see pages 10–11), but what about a smart factory? These are factories that employ intelligent robots to do the difficult and dangerous jobs that humans once had to do. The machines in these factories are all connected by the Industrial Internet of Things (see pages 8–9).

WHAT IF...? What if robots really can do a job more efficiently and accurately than humans – does that mean we should replace human workers? How might the workers feel about that? How might business owners who are trying to increase productivity feel?

THE INTERNET OF THINGS

The Internet of Things (IoT) is a giant network of devices that are connected to the internet and to each other. By gathering, sharing and analysing data, the smart devices that make up the IoT help to make our daily lives easier. Today, AI is being used to improve the way the IoT works.

SMARTENING UP

All over the world, billions of objects make up the Internet of Things. On their own, these are 'dumb devices', but linking them together so they can communicate with each other makes them smart! Many different types of device make up the IoT.

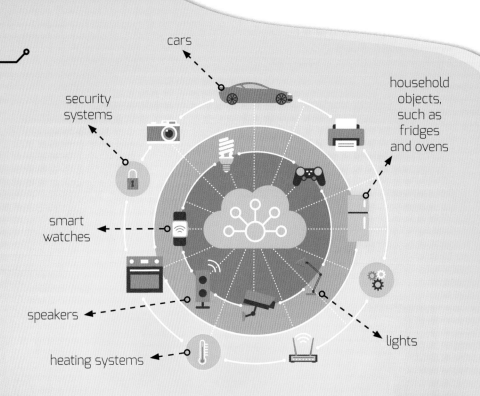

cars

household objects, such as fridges and ovens

security systems

smart watches

speakers

lights

heating systems

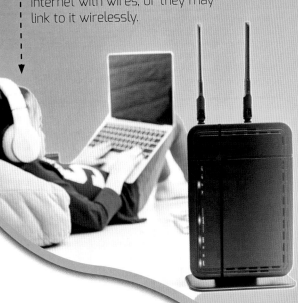

The devices on the Internet of Things may be connected to the internet with wires, or they may link to it wirelessly.

SENSORS AND SOFTWARE

The purpose of the IoT is to collect, analyse and share information. Devices then respond to that information to perform certain tasks. The smart devices connected to the IoT use sensors and **software** to act in an intelligent way.

 Sensors detect information from their environment (for example, temperature, light or movement).

 Software in the device looks at the data, analyses it and decides how to respond.

 An internet connection allows the data to be sent to other devices on the IoT. The device can also receive information from those devices to inform how it behaves.

WHAT'S THE USE?

Devices on the IoT can perform a whole range of tasks. For example, a sensor may detect when someone walks into a room, analyse that data and interpret it to mean that it needs to turn the lights on. A different device may respond to movement in a room by setting off an alarm. Data is collected, analysed and acted upon in a split second!

In 2020, there were an estimated 31 billion devices making up the Internet of Things.

Despite its name, the Internet of Things isn't the internet. It *uses* the internet to give computers and other devices the power to sense the world for themselves and communicate this information.

WELCOME TO YOUR SMART HOME

From lights that you can control remotely to curtains that close automatically at night, technology in the home is getting smarter and smarter. Some devices that seemed super intelligent just a few years ago are now so common that they're no longer considered AI. But new AI ideas are always coming along!

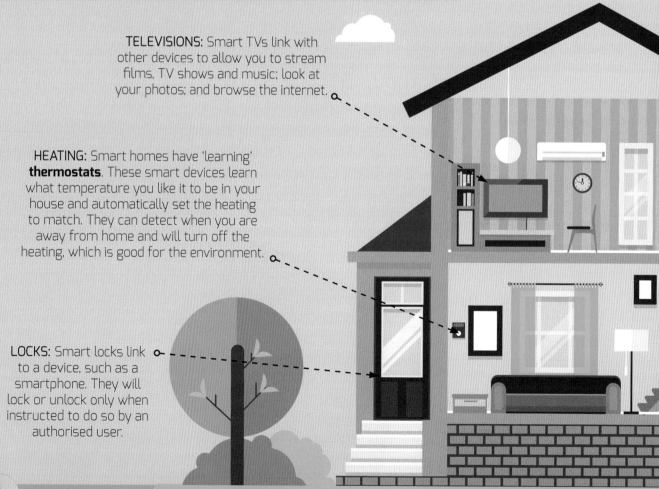

TELEVISIONS: Smart TVs link with other devices to allow you to stream films, TV shows and music; look at your photos; and browse the internet.

HEATING: Smart homes have 'learning' **thermostats**. These smart devices learn what temperature you like it to be in your house and automatically set the heating to match. They can detect when you are away from home and will turn off the heating, which is good for the environment.

LOCKS: Smart locks link to a device, such as a smartphone. They will lock or unlock only when instructed to do so by an authorised user.

Cocoon is an AI-operated 'home monitoring' security system. The speaker-sensor uses movement to learn its owner's daily routine. It uses sound and video technology to 'hear' and 'see' what's going on. If it detects anything out of the ordinary, it sends a warning message to the owner's phone.

Many devices in a smart home can be controlled remotely using a smartphone or a computer. Within the house, they can be operated by giving voice instructions to speakers with personal assistants, such as Alexa and Google Assistant (see pages 12–13).

SECURITY CAMERAS: Home security has never been better thanks to AI! The smartest homes might have cameras with **night vision** and sound that you can track on a remote computer.

SOUND SYSTEMS: Smart speakers can be placed all over the house and linked wirelessly. They can be used to play music, but also to send voice commands to other devices in a smart home.

LIGHTING: Smart lighting can be controlled by your voice or your phone. Smart light bulbs can change the brightness and colour of the lights in different rooms.

WHAT CAN I DO FOR YOU?

Every day, millions of people talk to a smart speaker – asking for information about everything from the weather, to the news, to when their parcel will be delivered. They probably don't even think about the incredible AI behind these interactions!

PERSONAL ASSISTANTS

Virtual personal assistants are basically AI software that can perform tasks based on instructions or questions that a user gives them. They can schedule meetings, place calls, give you the weather forecast – they can even tell a joke on demand! These assistants can also be used to control other devices in a smart home (see pages 10–11).

THE SOUND OF YOUR VOICE

Virtual assistants respond to voice commands. Human voices are unique, like fingerprints. We all have our own **tone**, **pitch** and patterns of speech. The smart software in a virtual assistant learns to recognise voice patterns. It also learns personal preferences so it can make recommendations, such as new music you might like.

Siri is the intelligent assistant on Apple devices like the iPhone.

Alexa is the virtual assistant software that comes with Amazon's smart speaker, the Echo. Microphones in the speaker detect the user's voice and send their command to a voice recognition service in the **cloud**. This interprets the instruction and sends it back to the device. Alexa can then answer the question or give instructions to other smart devices, from lights to door locks.

Virtual assistants are triggered by a 'wake word'. This is usually the name the assistant has been given by the manufacturer. To get a response from your Echo smart speaker, for example, you have to start your command or question with the wake word 'Alexa'.

THE POWER OF PERSONALITY

AI engineers think carefully about the personality of the virtual assistants they create. Teams of 'personality designers' work out what users might find most appealing. Do people want a 'human' feel to interactions with their speakers? Or do they want to keep a human/machine distance?

Try asking your smart speaker the questions opposite. What do the answers tell you? Why do you think the makers programmed the assistants in this way?

- Are you human?
- Are you a man or a woman?
- What's your favourite colour?

PERSONALITY

HUMANS...

Our personality is what makes us unique – it's all the mental characteristics that make us different from the person next to us. Our environment and experiences as we grow up affect our personality. In turn, our personality affects how we live our lives.

BEHAVIOUR: Are you the life and soul of the party? Or do you prefer not to do things that draw attention to yourself?

EMOTIONS: Do you have a positive, generally happy outlook on life? Or do you sometimes feel negative and sad?

REACTIONS: Are you a thinker, considering problems calmly? Or do you react quickly, going with your gut **instinct**?

RELATIONSHIPS: Are you outgoing, with a large social group? Or are you more shy, preferring just a few close friends?

Some people's personality and behaviour comes from what is considered to be socially acceptable.

...VS. MACHINES

BEHAVIOUR: Artificial intelligence could give machines behavioural characteristics, such as being chatty, bossy or helpful.

Personality is an important part of how we interact with the world around us. So, some people argue that if machines are going to be truly human-like – truly intelligent – they need to develop personalities. How do the features that inform human personalities apply to machines?

EMOTIONS: Intelligent machines can interpret data and decide what an appropriate, human-like emotional response would be, such as expressing excitement if a friend was coming for a sleepover.

RELATIONSHIPS: Being able to interact with humans and other machines is a key aim of AI. Understanding human language, responses and emotions is an important part of this.

REACTIONS: Machines don't have instincts or real emotions, so their reactions are fast and based directly on the data or information they receive.

Devices are not influenced by what other people think!

AI IN YOUR POCKET: THE SMARTPHONE

When mobile phones were launched in the 1980s, they changed the way people communicated forever. Phones have come a long way since then – from simple devices for making calls to truly intelligent machines that can perform all sorts of tasks.

INTELLIGENT PHONES

The first mobile phones were just that – devices that could make telephone calls without being fixed in one place. They soon began to offer more features, such as text messaging and address books. Things took another leap forward in 1994 when the first smartphone, the Simon Personal Communicator (SPC), was released.

The SPC was not smart by today's standards, but it had several features that are still used in phones, including a touch screen, calendar, address book and on-screen keyboard.

WHAT IF...? Some people believe that we already spend too much time on our devices and not enough time on real interactions. What if this get worse as technology gets better? Is it time we put down our gadgets and started talking to each other again?

INTERNET INTELLIGENCE

A lot of what smartphones can do is clever, but it isn't AI. Features such as email, game-playing and interacting with **apps** are possible because you can connect to the internet from wherever you are. But AI technology is becoming a bigger and bigger feature of mobile phones.

Voice assistants like the iPhone's Siri use speech recognition to respond to commands and allow you to control other smart devices.

The camera in a smartphone makes great use of AI. It identifies what type of photo you're snapping (a person or a landscape, for example) and can pick a filter that will bring out the best in your picture.

Smartphones include **facial recognition** – AI software that maps facial features. This can identify friends and family in your photos and on social media apps that you use via your phone.

Google's search engine is powered by AI technology, so when you search Google on your smartphone, AI operations kick in on the device.

AI IN ACTION

Some smartphones are now built with special AI microprocessors that can operate at super-speeds. These chips are sometimes called 'neural engines', because they can run special processes called **neural network algorithms**. They're really useful, because the faster your phone can perform an operation, the less battery it uses up!

17

WHERE DID I LEAVE MY KEYS?

Have you ever walked into a room and forgotten what you went in there for? Or been introduced to a new classmate and couldn't remember their name by lunchtime? Humans forget things. Machines like smart devices don't.

DAILY REMINDERS

One of the big benefits of smart devices is that they can remember things for us. They remind us when we have a dentist appointment and when our friends' birthdays are, for example. They store useful information that we couldn't keep in our own brains. All this is pretty smart, but it's based on information that we *give* our device, so it's not truly intelligent.

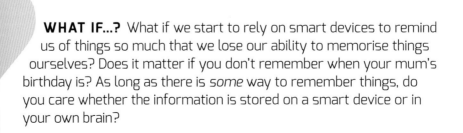

WHAT IF...? What if we start to rely on smart devices to remind us of things so much that we lose our ability to memorise things ourselves? Does it matter if you don't remember when your mum's birthday is? As long as there is *some* way to remember things, do you care whether the information is stored on a smart device or in your own brain?

SMART MEMORIES

As more AI is being put into smartphones, they need better memories. The memory also needs to be more adaptable to cope with the different types of information. AI tasks, such as facial recognition, image manipulation and **predictive text**, need special memory chips. Often, engineers take standard memory chips and add AI components to them to make devices smarter.

How many phone numbers do you know off by heart? Most people don't know any. But before smartphones stored all our contacts, people had to write down or learn their friends' numbers.

NEURAL NETWORKS

One of the key areas of AI is creating neural networks in computers. These networks are specially designed to work in a similar way to the human brain. Neural networks allow smart devices and other intelligent machines to remember and learn in the same way that humans can. For example, a device that can learn and remember the best route from A to B on an underground rail system will be using a type of neural network.

AI
IN ACTION

Existing random access memory (RAM) (see page 21) can struggle with complicated AI algorithms. Magnetic random access memory (MRAM) and resistive random access memory (RRAM) can cope with AI much better. Soon these types of memory may be used in smart devices in all homes and workplaces.

MEMORY

HUMANS...

What did you have for breakfast this morning? Do you know how your brain is recalling that information right at this second? 'Memory' refers to the process of acquiring, storing and retrieving information. This complicated process happens in three key stages.

1. ENCODING: A new memory is turned into a 'code' that your brain can process. This happens as chemical messengers in the brain jump the gap between nerve cells as electrical impulses.

2. STORING: You have both short-term and long-term memory. Short-term memory is quite limited. Experts think it may be able to store only seven items for up to 30 seconds. Some information moves into your long-term memory. There is no limit to the volume or time in long-term memory storage.

3. RETRIEVING: Retrieval, or recall, brings the stored memories back into your conscious mind. If you can't remember something properly it may be because it wasn't encoded properly (perhaps you weren't paying full attention).

Human memory never fills up, but over time the process becomes less efficient.

...VS. MACHINES

1. ENCODING:
Software in your smart device reads the information you give it as code, made up of numbers. This is how it understands your instructions.

Computer memory, such as the memory in a smart device, is also used for storing and retrieving information. But in the case of a computer, the memory centre of the 'brain' is the memory chip.

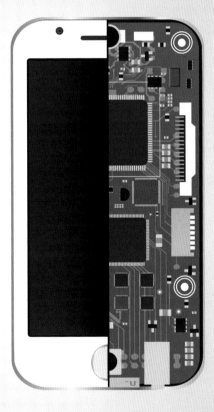

2. STORING: Computers have different types of memory, like humans. Random access memory (RAM) stores information temporarily, while it is needed. Read-only memory (ROM) is like long-term memory – it can store data for a long time, although there's a limit to how much it can store.

3. RETRIEVING: When a smart device is asked to retrieve a piece of information, such as a contact number, the request is sent to the memory chip, which pulls out the correct 'memory'.

Computer memory never deteriorates, but it can store only a certain amount of information.

HEALTH AND FITNESS

AI and smart devices have done a lot to change how efficiently and easily we get jobs done and go about our daily lives. But they have also had other benefits, especially when it comes to our health and wellbeing.

FITNESS TRACKERS

Smart devices, such as fitness trackers and smartwatches, mean we can now track every step we take. We can monitor the calories we eat and offset that against the exercise we take. Some trackers are smart enough to know whether we're walking, running, cycling or swimming just by the speed and type of movement they detect!

WHAT IF...? We're not the only ones who have access to the data logged on fitness trackers or apps. All that personal information could be seen and used by the company that made the device. What if we don't want to share that information? Has AI made it too easy to share personal data? Do we consider the consequences of this carefully enough?

We all know exercise is good for us, but it's not always easy to know if we're getting enough of it!

PERSONAL TRAINERS

AI can offer virtual personal trainers. AI apps will work out a programme designed just for you. If you're not meeting your daily goals, your smart device will send you an alert to let you know. And using your smartphone's camera, it will even be able to tell you if you're doing something wrong!

Fitness trackers range from simple devices that just count the steps you take, to high-tech watches that link with diet apps and your heath information.

Apple has created a special app to go with its smartwatch. The app uses sensors to take an electrocardiogram (ECG), which measures the heart's rhythm and electrical activity. This can give an early warning of heart problems, so users can get medical help.

AI
IN ACTION

WARNING SIGNS

Smart devices like fitness trackers can also track things like your heart rate and sleep patterns. Sensors in the smartwatch count your heartbeats. AI software analyses the data and sends an alarm if the information it receives suggests you are in danger. Features like this can really save lives.

WHY DID YOU DO THAT?

One of the reasons humans perform tasks so well is our ability to make logical decisions based on our understanding of the world. How can smart devices make the same type of logical interpretations in order to perform tasks as efficiently as we do?

POLANYI'S PARADOX

Try to explain to a friend how to ride a bike. Harder than it sounds isn't it? This is an example of 'Polanyi's **paradox**' – the idea that humans actually understand more than they are able to describe. This is explained as 'We can know more than we can tell'. So, if we can't explain what we know, how can we program devices to perform tasks?

MAKING JUDGEMENTS

AI engineers are finding a way around Polanyi's paradox by developing systems that have '**cognitive** skills'. These are the key skills your brain uses to think and solve problems – abilities such as learning, remembering and reasoning. Some cognitive tasks have proved pretty easy for machines to master, but others are harder to teach or recreate.

How can we program an intelligent robot to ride a bike if we can't describe that process clearly ourselves?

Playing chess requires cognitive skills, but brainy computer Deep Blue's AI technology helped it beat world chess champion Garry Kasparov in 1997.

READING BETWEEN THE LINES

Look at this statement:

> A woman went to a restaurant. She ordered pasta. She forgot to order a drink.

If I asked you you what the woman ate, you'd say 'pasta' wouldn't you? But look again – I didn't actually say that she *ate* pasta. You (correctly) **inferred** that from what I said. You *read between the lines*.

It's really difficult for computers, even intelligent ones, to do that. They need to be given precise information or data.

Common-sense knowledge is the area of AI that deals with programming computers with facts about the everyday world that humans just *know*. This covers all sorts of things, from 'sugar is sweet' to 'elephants don't drive cars'!

THE UNSOLVED PROBLEM

Programming computers with the 'rules' of common sense is a big task. In fact, experts haven't been completely successful in it so far – it's one of AI's unsolved problems. That's why when you ask a smart device a question, the answer doesn't always make sense!

COMMON SENSE

HUMANS...

Common sense refers to the way we react in a rational way to the world around us. The senses take in information, then our brain analyses it and decides how to respond. As well as common sense, instinct and 'gut feeling' also affect how we react to things.

MAKING INFERENCES: Your brain uses common sense to read between the lines. If you ask someone if they're okay, and they snap back 'I'm fine!' and stamp their foot, common sense tells you that they're actually *not* fine!

INTERPRETING LANGUAGE: Language is key to common sense. The knowledge we consider to be common sense is only 'common' if it can be communicated between people.

LEARNING FROM EXPERIENCE: Our senses teach us a lot. If you touch something hot, your brain registers the fact that it hurts. Common sense will stop you making the same mistake again.

SENSING THE ENVIRONMENT: We use our senses to see, smell, hear, touch and taste things in the world around us.

Common sense can be seen as one of the links between our five senses and our mind.

...VS. MACHINES

SENSING THE ENVIRONMENT: Sensors in smart devices allow them to 'read' the environment and react in a logical way (for example, a moving smart device will detect when it nears the edge of a surface and change direction to avoid falling off).

Machines, even intelligent ones, work in a different way. They don't operate on instinct – they need data and statistics in order to make decisions. But if common sense is defined as the ability to make good decisions and behave in a reasonable way, it seems that some machines do have common sense!

LEARNING FROM EXPERIENCE: 'Reinforcement algorithms' are a type of AI algorithm that helps machines learn through trial and error. By getting things wrong, they eventually learn to get them right.

INTERPRETING LANGUAGE: Speech recognition software already helps smart devices to respond logically to requests. Soon it may recognise tones of voice and learn to read between the lines of what people say.

MAKING INFERENCES: Some AI systems include an 'inference engine'. This applies logical rules to data, which allows it to work out new information.

Some devices can analyse external data and react in a 'common sense' way.

THE FUTURE OF SMART DEVICES

Smart devices are already a big part of our everyday lives.
But they are set to become even more important as AI evolves.
So, what might be the smart devices of the future?

SMARTER HOMES

Engineers are continually coming up with new intelligent devices for the home.

Smart microwaves will be able to automatically cook food for the right length of time.

Energy systems, such as heating and lighting, will warn you when you're going over budget.

Smart toilets may come with heated seats. They may even be able to tell if you are ill by analysing your pee and poo!

EVERYTHING'S CONNECTED

As more devices get smart and connect to the Internet of Things (see pages 8–9), our lives will become more efficient in various ways.

Imagine a car that can diagnose what's wrong with itself, communicate with the nearest garage and order a new part!

AI EVERYWHERE

Soon it will seem perfectly ordinary for AI to be built into homes, businesses and schools, in ways we wouldn't have been able to imagine even a few years ago. What smart device would you like to see become part of daily life in the future?

Imagine an alarm clock that will wake you earlier or later because it knows your train has been delayed, or that your journey is going to take longer because of bad weather.

WHAT IF...? Connectivity is key for smart devices. Without the internet, a lot of the things we rely on don't work. And the internet relies on electricity. So what if the internet goes down? Should we ensure there are old-fashioned back-ups, so that the basics we rely on continue to function? And if so, does that mean that smart devices aren't really necessary?

GLOSSARY

algorithm – a set of steps that tell a computer what to do in order to solve a problem or perform a task

app – an application designed for a mobile device, such as a smartphone or tablet

cloud – computer servers that can be accessed via the internet

cognitive – describes the process of gaining knowledge and understanding

cybernetics – the science of communications and automatic control systems in humans and machines

data – information, such as facts and statistics, that is collected and analysed

ethical – relating to whether things are right or wrong

facial recognition – software that can map out facial features in order to identify individuals in images

infer – to work something out using reasoning and evidence

input – information or data that is put into a machine or computer

instinct – knowing something or behaving automatically, without learning or thinking about it

neural networks – algorithms that recognise patterns and interpret data in a similar way to how the human brain works

night vision – technology that can 'see' in the dark

output – the information that comes out of a computer after the data has been processed

paradox – a statement or idea that seems difficult or impossible, but which may actually be true

pioneer – someone who is the first to explore, develop or use a particular idea or thing

pitch – how high or low your voice is

predictive text – a smartphone technology that suggests what word you may want to use before you have finished typing it

sensor – a device that detects and responds to physical things

software – the programs that give computers the instructions they need to work

thermostat – a device that automatically controls temperature by turning heating up or down

tone – the mood or attitude of your voice and speech

FIND OUT MORE

BOOKS

AI (The Tech Head Guide) by William Potter (Wayland, 2020)

Machine Learning (Explore AI) by Sonya Newland (Wayland, 2020)

The Internet of Things (Science for the Future) by Lisa J. Amstutz (Focus Readers, 2019)

WEBSITES

www.bbc.co.uk/newsround/49274918

Discover more about what AI is and what it does.

www.kidscodecs.com/what-is-internet-of-things

Learn about the Internet of Things and how smart devices can be connected.

INDEX